CENTURION
in action

by Stephen Tunbridge

illustrated by Don Greer

[Cover] An Israeli Centurion Mk 5, extensively modified, advancing along a road on the Golan Heights, during the Yom Kippur War, October 1973.

 squadron/signal publications

COPYRIGHT © 1976 by SQUADRON/SIGNAL PUBLICATIONS, INC.
1115 CROWLEY DR., CARROLLTON, TEXAS 75006

All rights reserved. No part of this publication may be reproduced, stored in a retrieval system, or transmitted in any form by any means electrical, mechanical or otherwise, without first seeking the written permission of the publisher.

If you have any photographs of the aircraft, armor, soldiers or ships of any nation, particularly wartime snapshots, why not share them with us and help make Squadron/Signal's books all the more interesting and complete in the future. Any photograph sent to us will be copied and the original returned. The donor will be fully credited for any photos used. Please send them to: Squadron/Signal Publications, Inc., 1115 Crowley Dr., Carrollton, TX 75006.

ISBN 0-89747-046-X

ACKNOWLEDGEMENTS

The Author is most grateful to all those governments, manufacturers and individuals who have supplied information and photographs, and in particular the RAC Tank Museum, the Imperial War Museum and particularly the Swiss and Australian Armies. Major Hiusur of the Danish Army and Captain Wallner of the Swedish Army, and Messrs. H.R. Kurz, C.F. Foss, D.P. Dyer, R. Surlemont, H.L. Doyle, A.J. Kaye, M.A. Roseberg and N. Ayliffe-Jones have been extremely helpful. Mention must also be made of the editorial service given by his father, Paul Tunbridge.

FORWARD

The Centurion remains in service with a number of armies. (Australia, Britain, Canada, Denmark, India, Iraq, Israel, Jordan, Kuwait, Lybia, Lebanon, Netherlands, New Zealand, South Africa, Sweden, Switzerland.) Extensive information has been supplied by most countries, while for others the barest details have been released, presumably for security reasons. For several variants of the Centurion, details were unobtainable. However, these variants are secondary and were not developed beyond the prototype stage; their absence, therefore, has not great significance.

A considerable amount of documentation and photographs have been released from British sources including the Imperial War Museum, the RAC Tank Museum at Bovington, etc. The full details of the early development of Centurion are still covered by the "thirty-years rule", apart from those documents released up to the end of World War II. Military historians and others will have to wait until 1976 before this material becomes available-for the present only records up to 1945 can be consulted.

Weights, dimensions, distances, etc., are given in metric or Anglo-Saxon units but for the convenience of the reader, it was decided to leave the original figures whenever possible, since these are round numbers and are therefore exact. For example, the 20 pdr is a well-known figure while its equivalent in millimetres (83.4 mm) would be unrecognizable. Similarly, the armour penetration of guns is always given, for instance, as 1,000 yards and not 914 m. On the other hand, specifications are always dimensioned in metric figures.

ABBREVIATIONS

AEC	Associate Equipment Company Ltd [Southall, Middlesex]
AOP	Armoured Observation Post
APDS	Armour Piercing Discarding Sabot
ARK	Armoured Ramp Carrier
ARV	Armoured Recovery Vehicle
AVLB	Armoured Vehicle Launched Bridge
AVRE	Armoured Vehicle Royal Engineers
BAOR	British Army Of the Rhine
BARV	Beach Armoured Recovery Vehicle
CDL	Canal Defence Light
DD	Duplex Drive
FV	Fighting Vehicle
FVRDE	Fighting Vehicle Research and Development Establishment
HE	High Explosive
HESH	High Explosive Squash Head
IR	Infra-Red
IDF	Israeli Defence Forces
MG	Machine Gun
Mk	Mark
MVEE	Military Vehicles and Engineering Establishment
pdr	pounder [gun calibre]
Pz	Panzer [in Swiss usage]
QF	quick firing
RAAC	Royal Australian Armoured Corps
REME	Royal Electrical and Mechanical Engineers
RMG	Ranging Machine Gun
RPG	Rocket Propelled Grenade
RTR	Royal Tank Regiment

[Right] Centurion Mk 3 of the British 8th Hussars Regiment in Korea crossing the Imjin River on an American-built pontoon bridge. The driver's goggles were essential protection against the thick dust thrown up on roads. Markings include a white star on front stowage box, a formation sign consisting of a black square with white circle [the sign of the 29th Brigade to which the 8th K.R.I.H. was attached] and the Royal Armoured Corps Arm of Service mark, the red over yellow square, with a black "41" as the Unit Serial. Barely visible is another Allied Star centrally located on the anti-bazooka plate. Almost obscured by dust is the Army Registry number in white on a black rectangle, on the lower front plate. [U.S. Army]

A-41 with 17 pdr and 20mm Polsten gun on left hand side of turret. Note that all optical equipment on the turret and for the driver has been removed. The first ten A-41 pilot models were all fitted with the 20mm Polsten gun, but this took up a disproportionately large amount of space and was considered too large for an anti-personnel weapon. [R.A.C. Tank Museum]

Centurion Mk I

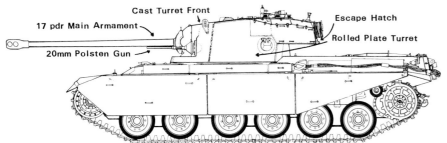

Centurion Development

During World War II the traditional British doctrine of having specialized categories of tanks to work with infantry on the one hand, and, to operate in armoured divisions against enemy tanks, on the other hand was revised as a result of experience gained in the North African campaigns. In September 1942, the War Office requirement for tank development laid emphasis on the need for a 'universal' or general purpose tank chassis which could be readily adapted to meet various specialized tasks including development in both the above-mentioned Infantry and Cruiser roles. In the past it had been necessary to have a number of chassis designs, each of a distinctive and separate pattern, with all the disadvantages inherent with this lack of standardization.

At the same time as design was in progress on the heavy 'Cruiser' A41 in 1944, work was also proceeding on the more heavily armoured 'Infantry' tank, the A45. The two tanks were to have a number of common assemblies in line with the Tank Board's decision of 1942 which had laid emphasis on the advantages to be obtained from standardization of designs for the two categories. It was intended that the weight of the A45 prototypes would be 55 tons with a maximum speed of 18 mph. These prototypes, which were to have a frontal armour of 6 inches equivalent thickness and incorporate the A41 turret and gun, were scheduled to be completed by mid-1946.

But by this time, the Sherman and the Churchill had demonstrated their ability of fulfilling the 'Cruiser' and 'Infantry' roles which had hitherto called for individual particular-purpose vehicles. This led to the abandonment of the separate Cruiser/Infantry concepts as a result of which only the A41 design which successfully met all the requirements for a 'universal' tank was proceeded with. Several years were to elapse before the A45 appeared in quite a different form as the FV 214 Conqueror.

Following operational experience in the European theatre where armoured formations had advanced 450 miles in nine days, Field-Marshal Montgomery

A41 PILOT MODELS

Pilot models	Designation	Main armament	Secondary armament	Distinguishing features
1-5	A41	17 pdr	20 mm Polsten	rear circular escape door
6-10	A41	17 pdr	20 mm Polsten - Optional linkage	rear circular escape door
11-15	A41	17 pdr	7.92 mm Besa MG - Optional linkage	rear circular escape door
16-18	A41S	77 mm	7.92 mm Besa MG - Optional linkage	7.92 mm Besa MG in ball mounting
19-20	A41S	77 mm	7.92 mm Besa MG - Optional linkage	rear circular escape door

had called for a Capital tank which was defined in his memorandum on British Armour, issued under a Top Secret classification on 21st February 1945.

The go-ahead was given in July 1943 for work to proceed on the development of a 'heavy cruiser tank'. By this time, and in the light of further operational use of armour, the General Staff had modified these priorities and although reliability was still the number one requirement, other factors had to be given attention. The priorities had now become:

1.) Reliability
2.) Durability (minimum running life of 3,000 miles)
3.) Maximum weight 40 tons
   Maximum width 10 ft. 8 in. (width of a Bailey bridge)
4.) Armament
5.) Armour
6.) Speed and endurance
7.) Adequate fighting compartment

Later that year it was decided that power was to be provided by the Meteor engine derived from the Rolls-Royce Merlin aero-engine that had more than proved its worth in RAF aircraft.

A.E.C. were appointed 'production parents' for the A41 project as it was now officially designated. The first mention of the A41 was when the outline specification was presented to the Tank Board in November 1943. All the knowhow derived from allied and enemy intelligence reports as well as user reports were incorporated in the A41 design.

Cross-country performance comparable with that of the Comet was accorded higher priority than road speed in this their first design by the Department of Tank Design which also stipulated the importance of a high reverse gear. At this time the German 88 mm gun was recognized as being a formidable weapon when used against tanks and had more than proved itself in the Western Desert fighting. It was therefore logical that any future tank protection should be proof against this 88 mm gun. As well as being able to deal with the Tiger tank, the main armament was required to fire a HE round.

The frontal armour specified for the vehicle was to be based on an equivalent thickness of 4 inches, this value being reduced to 60% for the sides. The A41 was also designed to give adequate protection to the suspension-one of the most vulnerable parts of a tank-by the incorporation of armoured skirting plates to counter the hollow charge anti-tank weapon menace. In addition, mines which were getting bigger and more difficult to detect made it necessary for belly and suspension components to be strengthened. The maximum weight of 40 tons had been originally laid down by the General Staff for the following reasons:

a) size and strength of bridges
b) transport by rail
c) width of vehicle for traffic movement
d) reliability decreases as the weight increases

The General Staff had later to increase this weight to 60 tons to ensure adequate protection against the developments that had taken place in German anti-tank weapons.

The final A41 specification was accepted by the Tank Board in February 1944. The aim was for pilots and pre-production models to be produced towards the end of that year so that in the absence of difficulties small scale production might begin in the second quarter of 1945. The Tank Board recorded that 'this project had their full support and should proceed with all possible energy.'

By May 1944 an order had been placed with the Ministry of Supply for 20 prototypes having 17 pdr guns-in fact, on the last five prototypes 77 mm guns were installed-with different combinations of 20 mm Polsten guns and 7.92 mm MG's.

By April 1945, six prototypes A41 were being made ready for service in Germany, but arrived too late to see action. In the meantime, the first production version of the A41* or Centurion Mk 1, of which 100 were manufactured, was well in hand.

Centurion Mk I Specifications

General
Crew	4 - commander, driver, gunner, loader
Battle weight	48,700 kg
Dry weight	45,700 kg
Power-to-weight ratio	12.3 hp/tonne

Dimensions
Overall length of hull (without gun)	765 cm
Overall length (gun in travel lock)	775 cm
Overall width	275 cm
Overall height	280 cm
Ground clearance	50 cm

Trackwork
Length of track on ground	455 cm
Width over tracks	325 cm
Track centres	265 cm
Track width	60 cm
Track pitch	14 cm
No. of links per track	109

Armour
Turret	front	152 mm
	sides	133 mm
	top	80 mm
Hull	front	76 mm
	sides	50 mm
	rear	38 mm
	top	50 mm
	below	20 mm-30 mm

Performance
Maximum speed	34.5 km/h
Maximum cross-country speed	24 km/h
Road radius	96.5 km
Fording depth	145 cm
Vertical obstacle	90 cm
Trench	335 cm

Powerplant
Model	Meteor Mk IV (or Mk IVA)
Type	V 60°
Engine power	600 hp at 2,550 r.p.m. [640 hp at 2,550 rpm]
Maximum torque	195 kg/m at 1,500 r.p.m. [205 kg/m at 1600 rpm]
No. of cylinders	12
Bore	137 mm
Stroke	152 mm
Displacement	27 l.
Compression ratio	6:1 [7:1]

Charging set engine
Model	Morris
Type	side valve
No. of cylinders	4
Bore	57 mm
Stroke	90 mm
Displacement	918 c.c.
Engine power	20 hp at 2,500 r.p.m.

Fuel capacity 545 l.
No. of fuel tanks 2

Transmission
Model	Merritt-Brown
No. of gears	5 F/2 R
Turning radii	
1st gear	4.85 m
2nd gear	12.5 m
3rd gear	20 m
4th gear	32 m
5th gear	42.5 m

Main armament
Designation	Ordnance, QF 17 pdr Mk 6
Calibre	3in (76.2 mm)
Armour penetration	74 mm/60°/1,000 yards
Muzzle velocity	885 m/s
Overall length	470 cm
Weight of gun	890 kg
Weight of gun without breech mechanism	815 kg
Weight of barrel	728 kg
Weight of breech block	51 kg
Length of recoil	28.5 cm
Traverse	360°
Maximum elevation and depression	+20° -12°

Secondary armament
Model	Besa 7.92 mm MG
Weight	22 kg
Weight of barrel	7 kg
Overall length	110 cm
Rate of fire	750-850 rpm

Main ammunition stowage 70 rounds

Sighting and vision equipment:
- Commander: rotatable vision cupola incorporating eight episcopes which enabled the commander to obtain a 360° view of the terrain from inside the vehicle. Blade vane sight fitted on turret roof. Prismatic binoculars.
- Gunner: Periscope. Telescope sighting x 3 magnification. Traverse indicator situated to the right and slightly in rear of gunner's position.
- Loader: Periscope
- Driver: Two periscopes mounted in each of the driver's hatch doors.

Jordanian Centurion Mk 1. The Jordanian army, being trained and equipped by the British follows British marking and camouflage style. Note the formation sign and arm of service marks on front fenders. Camouflage is soft patches of red brown over dark yellow.

Centurion Mk 1 with 17 pdr in travelling position. Note circular emergency escape hatch located to the right rear vertical plate of the turret. Partially hidden behind the right fender support is the Royal Armoured Corps Centre Arm of Service insignia, a square, diagonally divided red over yellow topped with the words RAC Centre in black on a white band. [E.C.P.-Armees]

Mk I Secondary Armament

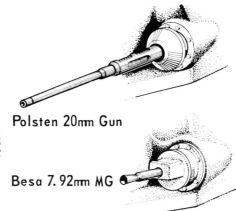

Polsten 20mm Gun

Besa 7.92mm MG

Rear view of Centurion Mk 1 with 17 pdr and 7.92 mm Besa MG. Marking on rear deflector plate is a convoy marking of alternating stripes of white and black. Note loading hatch on turret side. [RAC Tank Museum]

Centurion Mk 1 with 17 pdr and 7.92 mm Besa MG. Side plates are not yet fitted with hand grips on the example above. The upper tank is a test vehicle, used to show the feasibility of the up-armored Centurion Mk 2. Extra weights bolted to the glacis plate are used to measure strain on running gear since Centurion Mk 2 was heavier. Bottom vehicle also carries the 17 pdr and 7.92 mm Besa MG. [RAC Tank Museum]

SMOKE DISCHARGERS

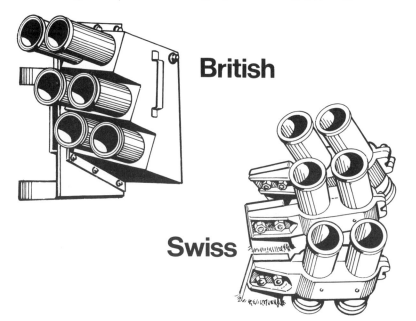

British

Swiss

Early A41 prototype fitted with CDL system. This equipment developed during the Second World War fitted to Matildas and Grants of the British Army was used for focusing light through a slit in the armour for night operations as an aid to shooting. [RAC Tank Museum]

[Above Right] Side-view of same vehicle. Although this project was discontinued, the CDL system was far less vulnerable to enemy reprisals than a conventional searchlight.

[Bottom] Centurion Mk 2 fitted with Schnorkel deepwading equipment. This type of equipment does not call for skirts like DD tanks, but the tank itself has to be 'sealed'. This tank is not able to swim as in the case of the Centurion DD. Rigid extensions are fitted for exhausts in addition to air intake and exhaust stacks. [RAC Tank Museum]

Centurion Mk 2

By January 1945, an up-armoured model, the A41A or Centurion Mk 2 was produced incorporating the first major modifications under the design parentage of Vickers-Armstrong at their Newscastle works. The Centurion Mk 2 was basically an up-armoured A41 hull with the difference that the final drive spur wheels were designed to give a different gear ratio (7.47:1 instead of 6.49:1) and the turret was formed from a one-piece casting which housed a rotatable vision cupola mounting nine episcopes and one periscopic binocular.

Instead of the Besa MG having a separate mounting in the turret, it was now mounted coaxially with the main armament, but remaining on its left. In this configuration the mounting enabled the Besa MG to have an elevation of 20° and a depression of 12°. Fitted in the turret roof for use by the gunner was a mounting and range gear for indirect laying with an incorporated periscopic sight for direct laying and observation. The main armament was stabilized in azimuth and elevation to ensure accurate shooting by neutralizing unsteadiness of the platform with the vehicle running, this was achieved by means of two electrically-driven gyroscopes for controlling the elevating and traversing gear respectively.

Centurion Mk 2

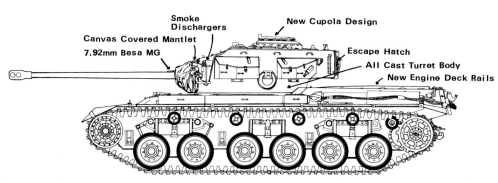

The A41A or Centurion Mk 2 was basically an up-armored A41 including such modifications as a new turret formed from a one-piece casting mounting a 17 pdr stabilized in azimuth and elevation. The 7.92 mm Besa MG is now mounted coaxially. The turret has a new rotatable vision commander's cupola and a centrally located rear turret escape hatch. [RAC Tank Museum]

Centurion Mk 3

By the summer of 1946, the Centurion Mk 2-the first of 100 mounting the 17 pdr gun-was running while the next 600 built to the same specification had the 20 pdr (83.4 mm) and were designated Mk 3. In December 1946 the British Army received its first Centurions Mk 1 and 2. By 1951 all Mk 2 models had been modified to Mk 3 some of which subsequently became Mk 5. The Mk 3 now had a 20 pdr developed by the Royal Armament Research and Development Establishment with a penetration of 96 mm/60°/1,000 yards instead of the 17 pdr which had a penetration of 74 mm/60°/1,000 yards. This new version was in most other respects similar to the Mk 2 but had an increased turret thickness of 25 mm. Furthermore, the length of the hull was decreased by 10 cm.

A Centurion Mk 4 designed for close support roles with the 95 mm howitzer as main armament and retaining the 7.92 mm Besa MG was abandoned in 1949 before reaching the production stage. If this proposed design had been accepted 10% of Centurion production would have been earmarked for this role.

Centurion Mk 3 with 20 pdr gun without fume extractor [type A barrel]. Other modifications included a 10 cm shortening in length of hull and rearrangement of air louvres on engine and transmission decks. [RAC Tank Museum]

Swedish Stridsvagn 81 [Centurion Mk 3] with type A barrel which later were modified along with Swedish Centurion Mk 5's [also designated Stridsvagn 81] by the installation of 105 mm gun. They were then designated Stridsvagn 102. [Swedish Army]

Browning M-1919 .30 Cal. MG

Centurion Mk 3

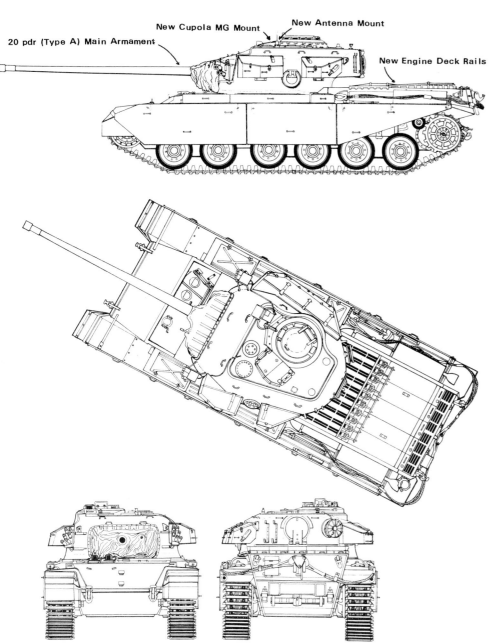

- 20 pdr (Type A) Main Armament
- New Cupola MG Mount
- New Antenna Mount
- New Engine Deck Rails

A New Zealand Army Centurion Mk 3 with 20 pdr [type A barrel] on training exercise.

Swiss Panzer 55 [Centurion Mk 3] fitted with 20 pdr gun with fume extractor [type B barrel]. Centurions in Switzerland incorporated such modifications as radio installations, smoke dischargers and MGs, all of Swiss manufacture. This early example is lacking MG and retains British style smoke dischargers. [Arthur Baur]

Early Swiss Panzer 55 [Centurion Mk 3] on training exercise. The tank has assumed a hull down position behind a small ridge. Sideskirts have been removed as they frequently became packed with mud and turf.

Swiss Panzer 55 [Centurion Mk 3] equipped with the 20 pdr gun on tank gunnery range. These three tanks are equipped with Simfire which obviates the need to use main armament shells and enables crew to train in simulated combat conditions. [Solartron Schlumberger]

Centurion Mk 5

In 1952 Vickers-Armstrong, Elswick, became the design parents of the Centurion Mk 5. By the end of that year, Vickers alone had 11 million worth of Centurions on their order book. Elswick was to remain a centre for Centurion production until 1959. The Centurion Mk 5 was also manufactured in quantity by the Ministry of Supply tank arsenal at Farington near Preston, and by Leyland motors. In point of fact, the majority of Mk 3 were modified to Mk 5 pattern except for reshaping of the turret roof and the retention of its rear escape hatch.

The Mk 5 was based on a Mk 3 hull but the 20 pdr fitted to the Mk 3 had a type A barrel. (i.e. without fume extractor fitted to the barrel) with a stabilizing counter weight to compensate for the absence of a muzzle brake. As the Mk 5, however, had a type B barrel (i.e. with fume extractor) no counter weight was therefore fitted. Improved ammunition was introduced for the 20 pdr fitted on the Mk 5 giving a penetration of 101 mm/60°/1,000 yards. In line with NATO ammunition calibre standardization, a US .30 in. M1919A4 Browning MG was fitted in place of the British co-axially mounted Besa MG. A second .30 in. MG was fitted to the cupola for use by the commander. The 2 in. bombthrower was also deleted from the Mk 5 presumably for reasons of obsolescence while the turret roof was reshaped with suppression of the rear escape hatch in the turret. Finally, a modification in the form of an additional guide roller was introduced to reduce the possibility of track throwing in difficult conditions.

Australian Centurion Mk 5 originally purchased as a Centurion Mk 3 and issued to the 1st Armoured Regiment in the early fifties. The Centurion Mk 3 later went through a modification program to bring it up to Mk 5 standards. This is why this Centurion Mk 5 retains the 20 pdr with type A barrel [i.e. without fume extractor]. Note British style Arm of Service Mark, red over yellow with a white stenciled unit serial. Also there is a tac sign partially hidden by the left side smoke dischargers. [Australian Army]

Centurion Mk 5

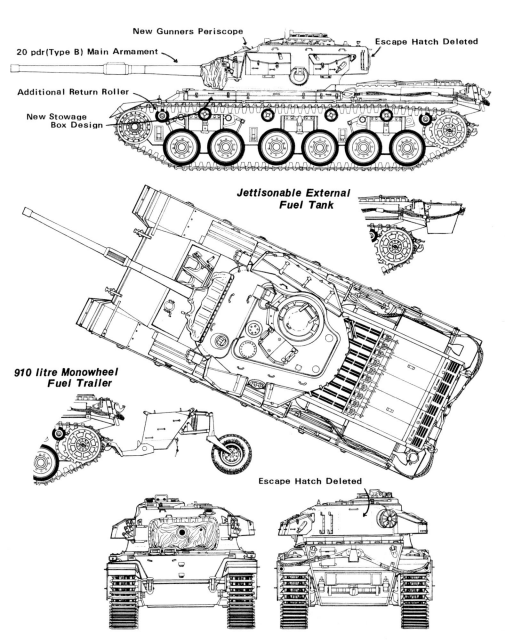

British Army Centurion Mk 5 which is distinguishable by the 20 pdr gun with fume extractor [type B barrel] and a .30 in. M1919A4 Browning MG in place of the British co-axially mounted Besa MG. The rear escape hatch in turret has been removed. [RAC Tank Museum]

Centurion Mk 7

Leyland Motors supervised the redesign of the Mk 5 hull which with additional fuel capacity and incorporating the Mk 5 turret became the Centurion Mk 7. This version made its appearance in 1953.

Following complaints from the RTR of the comparatively short range of the Centurion at about this time, jettisonable external fuel tanks were installed at the rear of the hull. A monowheel armoured fuel trailer with a 910 litre capacity later replaced the external fuel tanks. This trailer could be fitted by means of two hooks to the rear of the Centurion. When required, the trailer could be discarded by firing two small explosive charges by which the hooks were attached. Fuel was supplied to the tank engine by flexible hoses connected to two electric pumps. Once the initial driving experience was gained there were no problems in operating with this trailer. This equipment is still in service with the Swedish Army. However, both the external fuel tanks and trailer were in turn superseded on the Mk 7 onwards by a third fuel tank located within the tank at the rear of the transmission compartment. This fuel tank increased the fuel capacity from 550 litres to 1036 litres and nearly doubled the cruising range of the vehicle. Provision was made on the Mk 6, which was derived from the Mk 5, for an additional fuel tank to be fitted at the rear of the vehicle but externally. Further modifications included improved driver's controls and an ammunition loading hatch in the left hand side of the fighting compartment as well as two large headlights of prefocus type working on the double-dipping system of headlight control.

The main modifications distinguishing the Centurion Mk 7 from the Centurion Mk 5 are clearly visible. These include extended fuel tank with increased capacity as well as rearrangement of air louvres on engine and transmission decks. [RAC Tank Museum]

Swiss Panzer 57 with type B barrel. The Centurion Mk 7 incorporated various modifications including improved ammunition stowage. [Swiss Army]

Centurion Mk 8

In July 1955, a new turret design was completed for the Mk 8 with a canvas covered mantlet resiliently mounted on metalastic bushes carried on the cradle axis pins. New fire control equipment with a new type of elevation gear was fitted and also a contra-rotating cupola designed to ensure that the commander's line-of-sight always remained on the selected target irrespective of the direction of rotation of the turret. When the line-of-sight and axis of the bore of the gun were coincident, a line-up lock mechanism released the contra-rotating gear from the cupola. The loss of target involved with rotation of the former cupola was thus obviated. Additional protection was provided for the commander by the fitting of two semi-circular outward opening hatch doors instead of the simple hinged circular type on earlier models.

Trials carried out in 1956 demonstrated that the additional weight due to uparmouring of the Mk 8 had not affected its performance, riding characteristics, range or maintenance requirements. These trials were of some importance since they influenced the uparmouring policy of the Mk 10.

The Centurion Mk 8 was followed in 1959 by the Mk 9 which was an uparmoured Mk 7 fitted for the first time with the Vickers L7A1 105 mm gun.

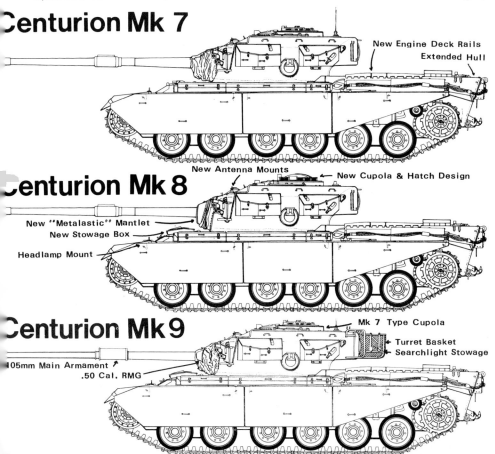

Centurion Mk 8 with 20 pdr gun and commander's cupola with two-part hatch. The headlights had previously been introduced on some Mk 7's. [RAC Tank Museum]

Basically a Centurion Mk 7, the Centurion Mk 9 was uparmoured and equipped with a 105 mm gun shown locked in travelling position. The presence of a turret basket and coaxial RMG and absence of IR night fighting equipment indicates a Mk 9/1. Note thermal sleeve surrounding the 105 mm gun-barrel. Note, also coaxial Ranging Machine Gun. [H.L. Doyle]

Centurion Mk 10

Improvements were made to the Mk 8, but this modified version was soon superseded by the Mk 10-an upgunned and uparmoured version of the Mk 8-which following development in 1960 first appeared the following year. The acceptance trials of the Mk 10-which was in many respects similar to the Mk 8-were successfully completed at the War Department Research Establishment in 1963. The design incorporated, however, a number of technical modifications to improve its performance. These changes mainly affected the armament and armour protection. In this way, a Vickers L7A1 105 mm gun-fitted retrospectively to most Centurions-mounted on impact resisting trunnions was incorporated in the construction in place of the 20 pdr. This gun with a penetration of 126 mm/60°/1,000 yards was able to fire APDS, HESH, canister and smoke rounds. A fume extractor was fitted to the barrel which did not incorporate a muzzle brake. A thermal sleeve eliminated gunnery errors brought about as a result of weather conditions such as wind, rain and sudden changes of temperature on the barrel. In 1960 a Centurion fired 17 rounds in one minute in the course of a demonstration with 100% hits on momentary targets. This was proof of the improved firepower of the upgunned Centurion with the 105 mm gun, the main armament of most Western medium-class tanks produced during the 1960s such as the Swedish 'S' tank, the West German Leopard, the Swiss Pz 61 and Pz 68, the American M60, etc. The new gun resulted in a notable increase in performance without calling for any major changes in design except for increased ammunition capacity and improved stowage. In this way, the Centurion Mk 10 was now able to carry a total of 70 rounds of ammunition of increased volume. A linked-sight graticule system for rapid alignment from commander to gunner was also fitted to the commander's cupola. Automatic stabilization of the gun control equipment was ensured by a velocity sensitive control system which came into operation at speeds over 4 m.p.h.

In order to keep abreast of modern technology and to increase the likelihood of first-round hits over long ranges, it was decided to install on Centurion Mk 10/2 onwards, with retrospective embodiment on Mk 6/2 and 9/2, a .50 in. RMG coaxially with the main armament and the .30 MG.

Centurion Mk 10/1 up to the latest Mk 13, included IR night fighting equipment (IR searchlight and headlights) and stowage baskets arranged round the turret. Centurion Mk 6/1, 6/2, 9/1 and 9/2 subsequently were also retrospectively fitted with IR equipment.

Centurion Mk 10 with turret basket and 105 mm gun protected by thermal sleeve. [RAC Tank Museum]

Centurion Mk 10

[Above] Centurion Mk 10 with 105 mm gun. This particular tank lacks the thermal sleeve round gun barrel. Similar variants were exported to Sweden where they were designated Stridsvagn 101. [RAC Tank Museum]

Centurions in Sweden underwent various modifications such as new Swedish 8mm MGs, American RA 421 radio installations, reinforcement of suspension for Swedish rough terrain, etc. In all, 110 changes have been made. In September 1974, the Swedish Government decided to maintain its 300 Centurions in service, but with further improvements to be made to existing Centurions to enable their Centurions to remain in service until well into the 1980's. [Swedish Army]

Centurion Mk 11 with its IR night fighting equipment, turret basket, one piece cupola hatch for commander, and coaxial RMG. [RAC Tank Museum]

[Above, Below, Below Left] Centurion Mk 12 easily recognizable with its IR night fighting equipment [searchlight and head-lights]. [RAC Tank Museum]

Centurion Specifications

	Mark 3	Mark 7	Mark 10
General			
Crew	4	4	4
Battle weight [kg]	50 813	50 813	51 616
Ground pressure [kg/cm²]	0.9	0.9	0.93
Power-to-weight ratio [hp/tonne]	12.5	12.5	12.5
Dimensions			
Overall length [cm]	983	985	985
Overall length [gun in travel lock] [cm]	861	861	861
Hull length [cm]	755	782	782
Overall width with side plates [cm]	338	339	339
Overall width w/o side plates [cm]	328	328	328
Height to top of cupola [cm]	294	302	301
Ground clearance [cm]	51	51	51
Trackwork			
Track width [mm]	609.6	609.6	609.6
Track pitch [mm]	139.7	139.7	139.7
No. of track links [dry pin]	108	108	108
Performance			
Maximum road speed [km/h]	34.6	34.6	34.6
Vertical obstacle [cm]	91	91	91
Maximum gradient [%]	60	60	60
Maximum trench [cm]	335	335	335
Fording depth [cm]	145	145	145
Cross-country range [km]	52.3	95.9	95.9
Road range [km]	100.5	184.4	184.4
Fuel consumption cross-country [l/km]	10.5	10.5	10.5
Fuel capacity			
Right-hand tank [l]	268.2	254.6	254.6
Left-hand tank [l]	281.8	350	350
Rear tank [l]	—	431.8	431.8
Powerplant			
Engine model	Meteor Mk IVB	Meteor Mk IVB	Meteor Mk IVB or IVB/1
Engine type	V 60°	V 60°	V 60°
Engine power [hp/rpm]	650/2550	650/2550	650/2550
Maximum torque at 1 600 rpm [kg/m]	211	211	211
Main cooling medium/capacity [l]	water/150	water/150	water/150
Fuel	petrol	petrol	petrol
Number of cylinders	12	12	12
Displacement [l]	27	27	27
Bore [mm]	137	137	137
Stroke [mm]	152	152	152
Compression ratio	7:1	7:1	7:1
Charging set engine			
Engine model	Morris USHNM	Morris USHNM	Morris USHNM
Engine type	side valve	side valve	side valve
Engine power [hp/rpm]	15/2500	15/2500	15/2500
Maximum torque at 1400 rpm [kg/cm]	455	455	455
Fuel	petrol	petrol	petrol
No. of cylinders	4	4	4
Displacement [c.c.]	918	918	918
Bore [mm]	57	57	57
Stroke [mm]	90	90	90
Compression ratio	6.7:1	6.7:1	6.7:1
Main cooling medium	water	water	water
Transmission			
Model	Merritt-Brown	Merritt-Brown	Merritt-Brown
No. of gears [F/R]	5/2	5/2	5/2
Steering system	triple differential	triple differential	triple differential
Clutch model	Borg and Beck	Borg and Beck	Borg and Beck
Clutch type	mechanical dry plate	mechanical dry plate	mechanical dry plate
Final drive ratio	7.47:1	7.47:1	7.47:1
Gear reduction ratio/vehicle speed [km/h]/turning radius [m]			
1st gear	11.643/4/4.9	11.643/4/4.9	11.643/4/4.9
2nd gear	4.593/10.1/12.2	4.593/10.1/12.2	4.593/10.1/12.2
3rd gear	2.855/16.3/20.1	2.855/16.3/20.1	2.855/16.3/20.1
4th gear	1.807/21.1/31.7	1.807/21.1/31.7	1.807/21.1/31.7
5th gear	1.343/34.6/42.7	1.343/34.6/42.7	1.343/34.6/42.7
Low reverse	22.894/2/2.4	22.894/2/2.4	22.894/2/2.4
High reverse	3.859/11.9/14.9	3.859/11.9/14.9	3.859/11.9/14.9
Electrical system			
Type	24 V nominal negative ground	24 V nominal negative ground	24 V nominal negative ground
Battery	4 off	4 off	4 off
Voltage [V]	6	6	6
Capacity [Ah]	150	115	115
Main engine generator			
Rating	55A, 27V at 1450 rpm	55A, 27V at 1450 rpm	55A, 27V at 1450 rpm
Charging set engine generator	120 A, 27V at 1450 rpm	120A, 27V at 1450 rpm	120A, 27V at 1450 rpm
Ammunition			
Main armament	65	63	70
Secondary armament MG	3600	4500	4500
RMG	—	—	—

Centurion Mk 13

The Centurion Mk 13 represented the ultimate in Centurion development. Main features are 105 mm gun with thermal sleeve, RMG, two-piece cupola hatch for commander, turret basket and IR night fighting equipment. Centurion Mk 13 has its coaxial Browning MG positioned higher than in Mk 12. This tank is painted in the most recent British Army camouflage, alternating many stripes of olive green and black. [RAC Tank Museum]

Three more views of the Mk 13, the final production version of the Centurion. [RAC Tank Museum]

Centurion Development

Basic Mark	Designation	Derived from	State	Main armament	Secondary armament	RMG equipment	IR equipment
Mark 1	A41*	A41	—	17 pdr	One 7.92 mm Besa MG	—	—
Mark 2	A41A	Mark 1	uparmoured	17 pdr	One 7.92 mm Besa MG	—	—
Mark 3	—	Mark 2	—	20 pdr	One 7.92 mm Besa MG	—	—
Mark 4	A41T	—	—	95 mm	One 7.92 mm Besa MG	—	—
Mark 5	FV 4011	Mark 3 hull	—	20 pdr	Two .30 Browning MG	—	—
Mark 5/1	FV 4011	Mark 5	uparmoured	20 pdr	Two .30 Browning MG	—	—
Mark 5/2	FV 4011	Mark 5	—	105 mm	Two .30 Browning MG	—	—
Mark 6	FV 4011	Mark 5	uparmoured	105 mm	Two .30 Browning MG	—	—
Mark 6/1	FV 4011	Mark 6	—	105 mm	Two .30 Browning MG	—	IR night fighting equipment
Mark 6/2	FV 4011	Mark 6	—	105 mm	Two .30 Browning MG	.50 RMG	IR night fighting equipment
Mark 7	FV 4007	New hull with Mk 5 turret	—	20 pdr	Two .30 Browning MG	—	—
Mark 7/1	FV 4012	Mark 7	uparmoured	20 pdr	Two .30 Browning MG	—	—
Mark 7/2	FV 4012	Mark 7	—	105 mm	Two .30 Browning MG	—	—
Mark 8	FV 4012	Mark 7 hull	—	20 pdr	Two .30 Browning MG	—	—
Mark 8/1	FV 4012	Mark 8	uparmoured	20 pdr	Two .30 Browning MG	—	—
Mark 8/2	FV 4012	Mark 8	—	105 mm	Two .30 Browning MG	—	—
Mark 9	FV 4015	Mark 7	uparmoured	105 mm	Two .30 Browning MG	—	—
Mark 9/1	FV 4015	Mark 9	—	105 mm	Two .30 Browning MG	—	IR night fighting equipment
Mark 9/2	FV 4015	Mark 9	—	105 mm	Two .30 Browning MG	.50 RMG	IR night fighting equipment
Mark 10	FV 4017	Mark 8	uparmoured	105 mm	Two .30 Browning MG	—	—
Mark 10/1	FV 4017	Mark 10	—	105 mm	Two .30 Browning MG	—	IR night fighting equipment
Mark 10/2	FV 4017	Mark 10	—	105 mm	Two .30 Browning MG	.50 RMG	IR night fighting equipment
Mark 11	FV 4017	Mark 6	—	105 mm	Two .30 Browning MG	.50 RMG	IR night fighting equipment
Mark 12	FV 4017	Mark 9	—	105 mm	Two .30 Browning MG	.50 RMG	IR night fighting equipment
Mark 13	FV 4017	Mark 10	—	105 mm	Two .30 Browning MG	.50 RMG	IR night fighting equipment

Centurion Mk 13

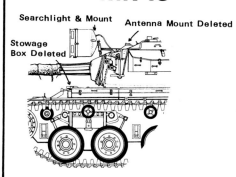

Specialized Versions of Centurion

Most tanks developed in the past have formed the basis for a family of variants designed for specialized roles such as bridgelaying, recovery, etc. Such has been the case with the Centurion whose basic design has lent itself to about twenty adaptations some of which progressed to the production stage. In this way, Centurion ARK, AVRE, BARV, ARV, and Armoured Bridgelayer all saw service with the British Army, the latter two versions being exported in large numbers to almost every user of the Centurion. In addition, a hydraulic dozer blade kit was also purchased by overseas buyers.

A number of developments of Centurion did not reach production. Worthy of mention were FV 3802 and FV 3805 (self-propelled guns) and FV 4004 and FV 4005 (tank destroyers). Some Centurions were equipped experimentally with a flame gun unit, or CDL system, or even Swingfire anti-tank missiles. Other Centurions were used in research development of other tanks and designated FV 214 Caernavon for the Conqueror and FV 4202 for the Chieftain. Equipment was developed to give deep-wading capability, but many of these systems did not go beyond prototype stage. Finally, Vickers in England, the Israeli Army and the Netherlands Army have each developed modernized versions of Centurion. As well as modified Centurion Bridgelayers introduced by the Netherlands Army, prototype designs have been completed by the Italian plant Astra.

Centurion AVRE Mk 5 designated FV 4003 carrying a section of metal roadway on cradle instead of the usual fascine bundle. This fascine could be jettisoned from its cradle by electrical firing of blow-out pins. [C.F. Foss]

Centurion AVRE Mk 5 towing the specially designed 15-ton two-wheel trailer to which is fitted the Giant Viper mine-clearing explosive equipment. The Giant Viper first developed during the Second World War was fitted to the Churchill AVRE. This vehicle was replaced in 1965 by the Centurion AVRE Mk 5, but this trailer as well as the Giant Viper remained in service, this time with the Centurion AVRE Mk 5. Before reaching the minefield, the Giant Viper rocket is fired taking with it a long length of explosive-filled hose which, slowed by parachute, falls on to the minefield and detonates. [C.F. Foss]

Centurion AVRE Mk 5 towing the special two-wheel trailer used for carrying the Giant Viper. The Centurion AVRE operates in battle conditions for the demolition of strong points or buildings using a low-velocity charge from the BL 165 mm L9A1 gun. The hydraulically-operated dozer blade clears debris and obstacles and is used for rough levelling and ditching. Note the bridge classification circle on the dozer blade which is yellow with black numeral. [H.L. Doyle]

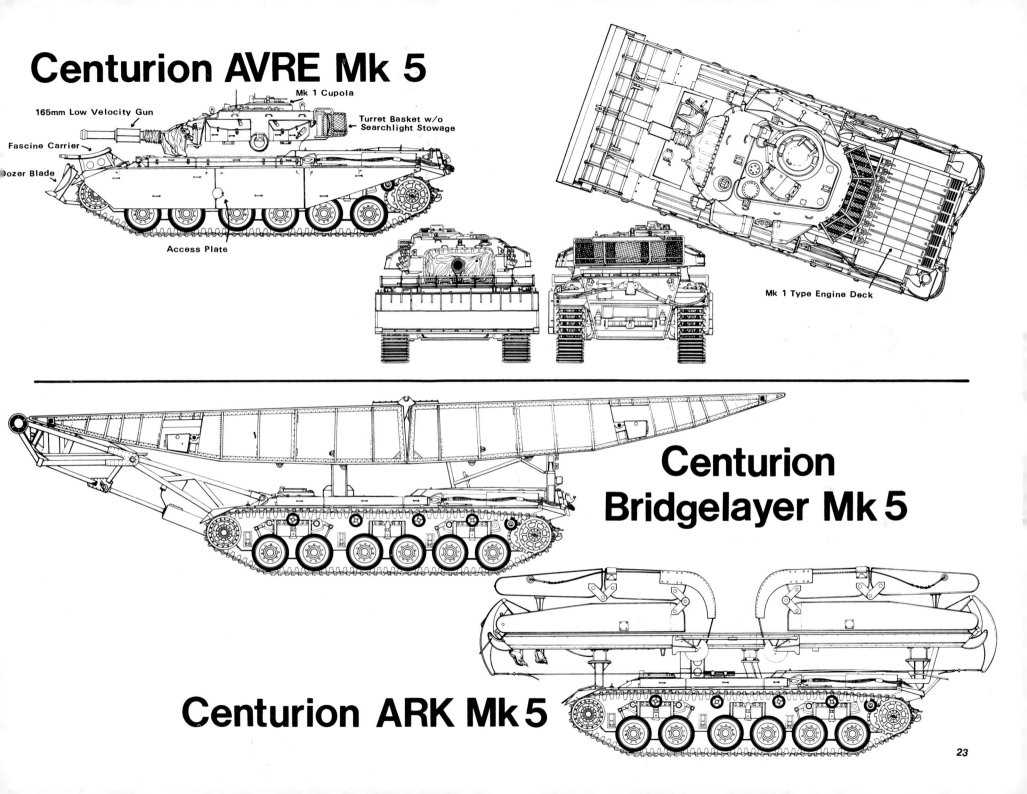

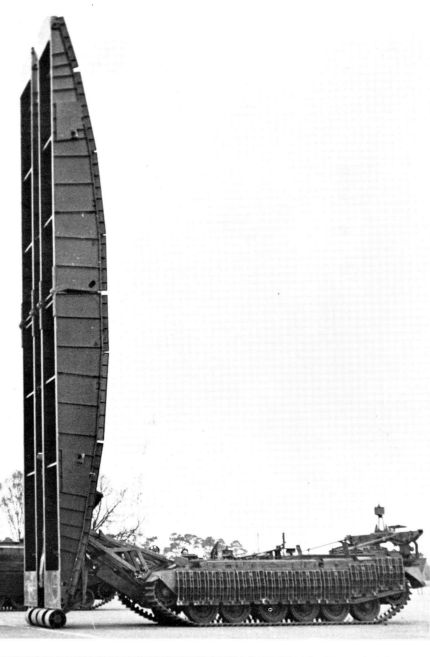

Photo sequence showing Centurion Armoured Bridgelayer Mk 5 launching bridge. The complete operations takes only 100 seconds. One important feature of the bridge is that once laid it can be picked up again on the other side of the gap. The 80 cm longitudinal gap in the bridge is filled when wheeled vehicles cross over by the insertion of deck sections normally carried on side of vehicle as shown. [C.F. Foss; H.L. Doyle]

[Below] Detail view of hydraulic launching arm of Centurion Bridgelayer Mk 5. The hydraulic arm rolls on the ground during launching or removal of bridge. [H.L Doyle]

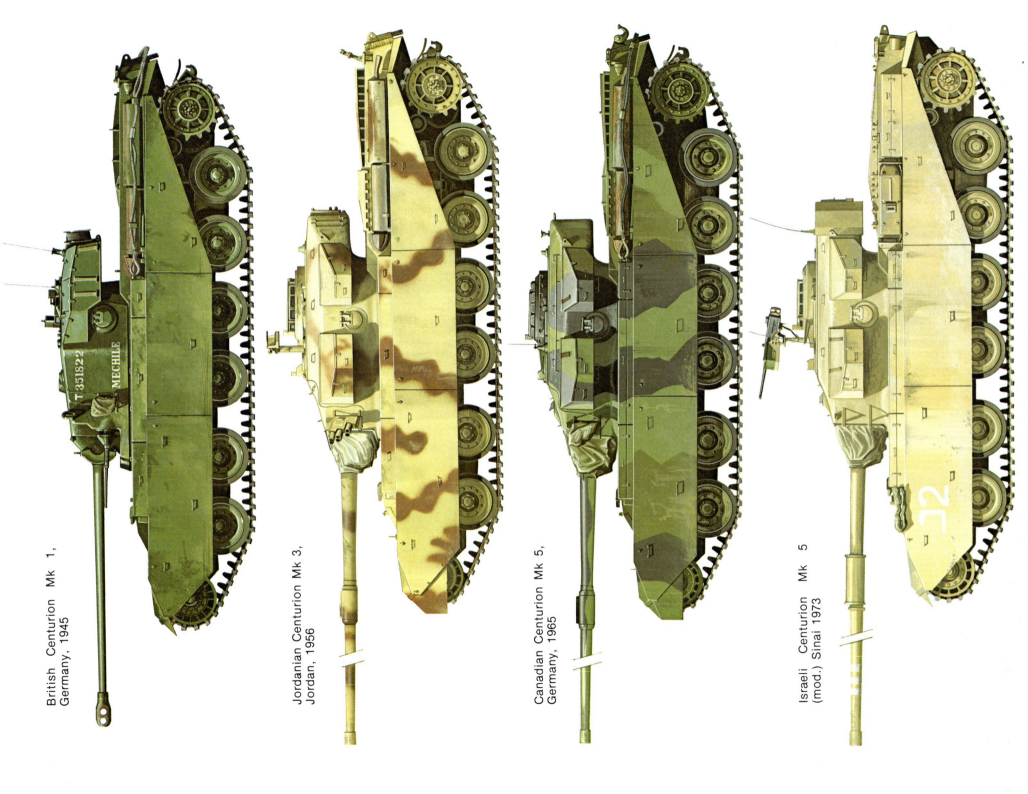

British Centurion Mk 1, Germany, 1945

Jordanian Centurion Mk 3, Jordan, 1956

Canadian Centurion Mk 5, Germany, 1965

Israeli Centurion Mk 5 (mod.) Sinai 1973

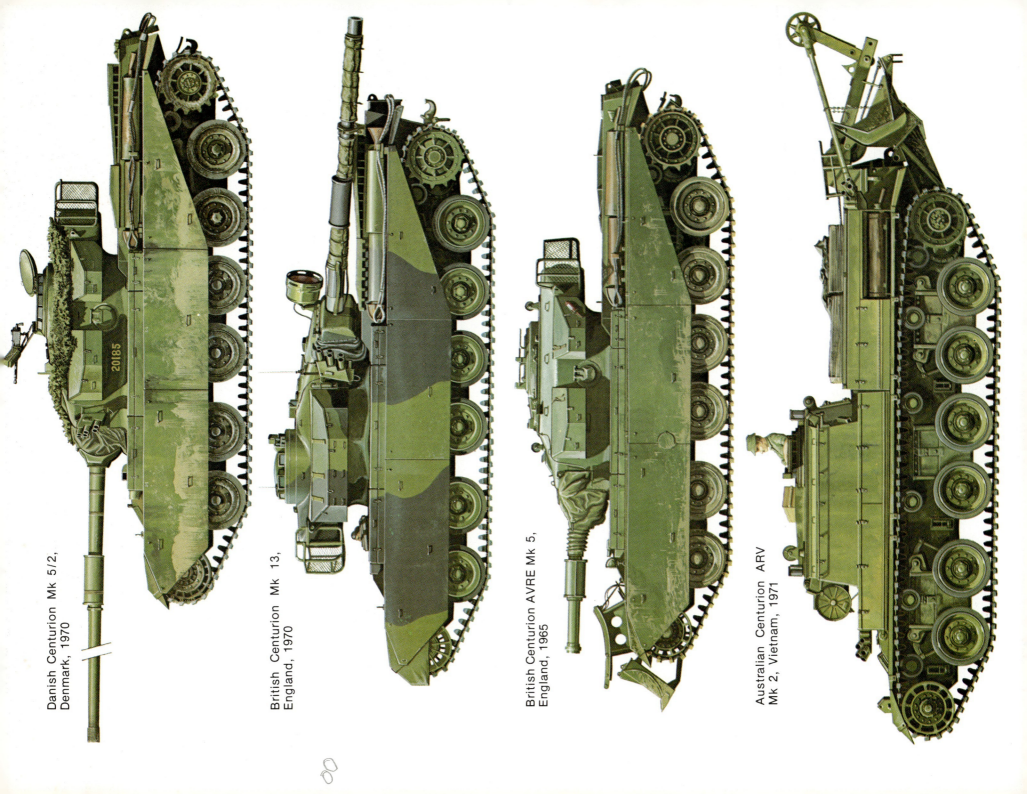

Centurion Bridgelayer and Centurion ARK compared. The ARK enters the river bed where its ramps are deployed on either side to bridge gap. In the second case, the ARK without ramps takes up position parallel with banks enabling the Centurion Bridgelayer to launch its first bridge. A second Bridgelayer launches the second bridge section, after positioning on first bridge. [RAC Tank Museum]

[Above Left] Rear-view of British Army Centurion ARK Mk 5. Basically a Centurion Mk 5 with turret removed, this vehicle originates from designs originally incorporated on Churchill and Sherman tank chassis during the Second World War. The extended pipe exhausts are employed for deep wading operations. [RAC Tank Museum]

[Below Left] British Army Centurion ARK Mk 5 at full speed. [C.F. Foss]

The Netherlands Army has adapted the AVLB bridge carried on the American M48 and M60 to their Centurions. The vehicle is seen here launching its bridge which makes an interesting comparison with the original Centurion Bridgelayer [FV 4002] with its one-piece bridge which is longer; 19.2 m against 13.7 m. [Netherlands Army]

British Army Centurion ARV Mk 2 on observation. The British Army was still operating these vehicles in 1975 pending replacement by the more sophisticated Chieftain ARV. Note front screen for driver. [C.F. Foss]

[Left Above] Entpannungspanzer 56 [Centurion ARV Mk 2] of the Swiss Army shows the multi-barrelled smoke dischargers and 7.5 mm 1951 model MG instead of the usual .30 in. Browning MG. The strengthened idler and sprocket adjustments are standard equipment on all Centurions. [Swiss Army]

Rear view of Swiss Entpannungspanzer 56 [Centurion ARV Mk 2] showing ground anchor allowing pulls to be taken in any direction. The 'A' frame is used for hauling unserviceable vehicles with a 30-ton vertical lifting capacity reduced to 12 tons when towing. Note wooden beams used for extricating tracked vehicles when bogged in mud or snow. [Swiss Army]

Centurion ARV Mk 2

Three-quarter front and rear views of Centurion BARV, designated FV 4018. This vehicle was mainly used for recovering 'drowned' equipment and bogged vehicles or nosing-off beached landing craft. A special rope-covered front bumper was used in this latter role. A heavy laminated glass screen made forward vision possible for the driver when working under water. [RAC School of Tank Technology]

Centurion modified by Vickers and Marconi to meet modern Main Battle Tank Standards. A 720 bhp diesel motor developed by General Motors, a TN12 gearbox as used in Chieftain, new ventilation, night vision equipment for commander and driver and modernized gun control and stabilization equipment have been installed. [Vickers]

Three-quarter front view of FV 3805. This 5.5 inch self-propelled gun was developed following cancellation of the FV 3802 development. The FV 3805 project was also cancelled as the 155 mm had been adopted as NATO standard calibre. Development continued with a totally different type of 105 mm self-propelled gun designated FV 433, better know as the Abbot. [Leyland]

Front view of the experimental FV 4004 better known as the Conway and based on a Centurion MK 3 hull. Pending introduction of the Conqueror, British armoured formations were to be equipped with this tank destroyer comparable to the Soviet JS3 fitted with a 122 mm gun. The vehicle is on display at the RAC Tank Museum, Bovington. Only one prototype was built with a four-man crew, driver, commander, gunner and loader, the latter three positioned in the turret. Armament is one 120 mm L1A1 gun and one .30 in. Browing MG. [R. Surlemont] [FVRDE via Armour School]

The A45 project approved in 1944 as an infantry tank was intended to be complementary to the A41 Heavy Cruiser Centurion and was chosen under the general designation for the FV 200 series from which was derived an FV 201 version equipped with 20 pdr gun. The FV 201 mock-up was completed in June 1947 and reached prototype stage but development was discontinued for technical and financial reasons. Illustration shows A45 equipped with Centurion turret mounting a 17 pdr gun instead of the 20 pdr supposed to equip production models. [Profile No 38-Conqueror]

"Abbots Pride", a Centurion Mk 3 of the 8th Hussars in vicinity of Imjin River. All Centurions sent to Korea were the Mk 3 with the 20 pdr guns type A barrel (i.e. without fume extractor). Most Centurions in Korea carried the United Nations white star recognition mark. Hanging tactical sign on removable metal plates was standard practice. [Imperial War Museum]

Centurion in action

Tank designs in the Second World War were influenced by those of the German Army which had proved so successful in the spectacular Blitzkrieg operations during the first two years of hostilities. The massive employment of armour on the Russian front and in North Africa resulted in soaring production figures and the introduction of more effective vehicles with vastly improved performance. From 1943, tanks were employed in close support of infantry in Italy, and later in the Far East, but towards the end of the war successes of armies were being measured in terms of advances of hundreds of miles a day.

It was in such a setting that was forged the Centurion project—one of the last tanks to reach prototype stage before the end of hostilities. By April 1945, six prototype A41 tanks, now referred to as Centurions, were being shipped for service in Germany to obtain battle and maintenance experience under combat conditions. These six tanks dispatched to 22nd Armoured Brigade of the 7th Armoured Divison under the code name **Operation Sentry** arrived too late to see action, the German Reich having accepted an unconditional surrender on 30th April 1945. The operation continued but as a routine troop trial under peacetime conditions.

It was 1950 before the Centurion gained battle experience in Korea. Although hostilities broke out on 24th June, Centurions did not leave England until the embarkation in October of the 8th King's Royal Irish Hussars forming part of the 29th Independent Brigade Group. The Centurion Mk 3 sent to Korea emerged as the best tank in the field even in comparison with the American M46 and the Soviet T-34/85. The Centurion contribution was invaluable, especially at the battle of the Imjin River where they covered the retreat of British troops faced by massive waves of Chinese forces.

Several years later, Centurions again saw action, but in a totally different theatre. In 1956, following the nationalization of the Suez canal, the United Nations was informed that armed forces of Israel had penetrated deeply into Egyptian territory. Britain and France, after having first called for a cease-fire, decided to take action against military targets in Egypt. It was decided to occupy Port Said to permit the landing of troops and equipment leading to the occupation of the Canal Zone and Cairo. Large forces had to be put together including 93 Centurion Mk 5s but shortly after the landing of the first Squadron of Centurions at Port Said, a cease-fire was declared before the Centurion was really blooded in battle.

Nearly a decade elapsed before Centurion guns were again fired in anger. In 1965, when the Kashmire issue flared up, the Indian 1st Armoured Division's Centurions demonstrated their outright superiortity over the Pakistani M4 Shermans, and comparatively modern M47 and M48 Pattons. Even in 1971, the Centurion Mk 5 and 7 of the Indian Army were still considered among the best tanks on the field, even in comparison with the T-59 of the Pakistani Army.

In the meantime, Centurion had twice experienced battle conditions in the Middle-East during the so-called "Six Day War" in June 1967 and again in October 1973. The Israeli Army was equipped with several hundred Centurions, the majority of which had been modernized and even in the face of the most powerful Soviet tank at that time, the formidable T-62, the Centurion proved highly successful.

The Centurion was also operational in South Vietnam in 1968 when units of the 1st Armoured Regiment went into action. As before, the Centurion Mk 5 proved its superiority over other tanks whether American or Soviet up to the time of its withdrawal in 1971.

"Carnoustie" a Centurion Mk 3 at full speed crossing a river with gun in rear travelling crutch. Tac sign is white shaded dark circle painted directly on anti-bazooka plate. Note absence of white star. [Imperial War Museum]

[Below Right] Centurion Mk 3 with crew taking up position as enemy movements are reported. This tank belongs to another unidentified unit, possibly the 1st British Commonwealth Division which was formed in July, 1951 from Commonwealth units in Korea. This division's formation sign was a blue shield with the Imperial crown, frequently having the word "Commonwealth" in gold on white underneath. Serial remained the same. Note also the nonstandard placement of the Registry number. [Imperial War Museum]

Rear view of a Centurion Mk 3 of the 8th Hussars after crossing the Imjin River. [Imperial War Museum]

Centurion Mk 3 of the 8th Hussars moves into a "dug-in" position on south bank of the Han River where Centurion first went into action against enemy armour. These were Chinese captured Cromwell tanks which were engaged at a range of 3,800 yards. Front markings have been repeated including tac sign on turret rear and convoy mark on deflector. The tank's name [on the side stowage box] is unclear, but may be "CALDERA". [U.S. Army]

Centurion Mk 3 of the 8th Hussars bogged down in an ice and snow-covered stream in Korea. Tanks were frequently stuck in paddy fields. "41" unit serial has been stencilled smaller than standard practice, also the white circle of the formation sign has been left broken as stencilled. Visible on the side of the tank is the tactical sign "3C" inside a circle, reflected in the water just below it. [U.S. Army]

Replenishing ammunition in Korea. Stowage was available for 65 rounds for the Mk 3's 20 pdr. main armament. [Imperial War Museum]

Tanks in Korea

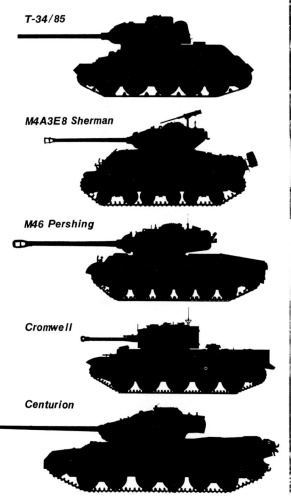

T-34/85

M4A3E8 Sherman

M46 Pershing

Cromwell

Centurion

The same vehicle after the crew has completed replenishment of ammunition. The tank is camouflaged since the distant hills are held by Chinese forces. The crew's helmets are slung over the side of the turret. [Imperial War Museum]

Jordanian Centurion Mk 5 with 20 pdr. gun. [Jordanian Army]

Canadian Centurion Mk 5 showing its 3-color dazzle camouflage. The most common color combination used by Canadians is yellowish green drab, olive drab and black. [Canadian Army]

An Israeli Centurion advancing at high speed during the Sinai Campaign in 1967. The two white rings on the 105 mm gun barrel indicate the second company of an armored battalion.

[Above Left] Israeli Centurion crews mount during a training exercise. This photo shows up gunned Mk 5's modified on a similar pattern as Dutch Centurions. The main armament is Vickers L7A1 105 MM gun. The cupola mounted MG is a U.S. made .50 cal. The Chevron marking, in white on an overall sand yellow tank, is a basic Israeli tactical marking. Barely visible on the central lower bow plate is army serial number, which is a six digit number followed by the Hebrew letter "tzen" in white on a black rectangle, the "tzen" being the first letter of Tzahal or army.

A rare example of number and letter tactical markings seen on Israeli Centurion Mk 5's. These combinations were commonly seen on Israeli M 48's and Super Shermans. In this case the marking on the lead tank is "Beth 2". That on the following tank appears to be the same. Note, however, that while the first vehicle has 3 white bands on the 105 mm barrel, the second has only one.
A41 with 17 pdr and 20 mm Polsten gun on left-hand side of turret. Note that all optical equipment on the turret and for the driver has been removed. The first ten A41 pilot models were all fitted with the 20 mm Polsten gun but this took up a disproportionately large amount of space and was considered too large for an antipersonnel gun. [RAC Tank Museum]

Two Centurions excort U.S. built M3A1 half tracks along a road on the Jordanian Front. The lead tank carries a single barrel ring and the downward pointing chevron, the tactical symbol of the III tank battalion. [Israeli Army]

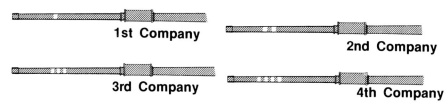

Two Mk 5's wind up a hill on the Golan Heights. The only marking they share are three white bands on the gun barrels. These were tactical markings and denoted the third company of the battalion.

Israeli Centurions advancing along the Joranian front during the 1967 campaign. Commanders who usually fought with their head out of the cupola suffered heavy losses. The commander of the Centurion on the left has opened his hatch at 90° to provide protection for his back. Side-plates were often removed since they proved a hindrance in battle conditions when tracks and road wheels had to be changed rapidly. Black and white striped square on fuel tank is a convoy marking symbol similar to the British. [Israeli Army]

Israeli Tactical Signs

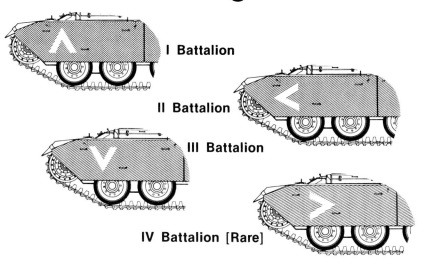

As a result of their extremely successful used armor in the 1956 War, there was a rapid expansion of the Israeli armor formations. By 1967, Tzahal armor units were almost completely standardized into Brigades which, in turn, were composed of three battalions, each made up of four companies. This system was continued during the Yom Kippur War of 1973. But due to rapid, piecemeal mobilization many photos of 1973 War show tank groups with widely varied markings. The confusion was, of course, compounded by repaired tanks being returned to whichever unit could grab them first.

Jordanian Centurions abandoned in Bethlehem in 1967, carrying standard British desert camouflage of red brown patches over dark yellow. The center tank is a Mk 3 with the type A 20 pdr, the rest appear to be Mk 5's. The Israeli intention to press them into service is obvious from the hand applied Star of David added to the front of the second vehicle from the left.

Six Israeli Centurions knocked out in the Golan Heights fighting during the Yom Kippur War of 1973. Chevrons are in use as in the 1967 war, and indicate tanks from at least two different battalions.

Two Israeli Centurions resting inside Syrian Territory. A third is barely visible coming over the ridge in the background. The vehicle on the left has a forward pointing chevron almost obscured by a heavy layer of dust. The tank on the right carries a tactical designation [Aleph One] painted in black on an extra stowage box on the left rear of the turret that was a usual Israeli modification.

Swiss Panzer 55/60 [Centurion Mk 3] equipped with Vickers licence-built 105 mm gun [model 1960]. The crew is removing track while tank commander directs driver's movements. Later Swiss Centurions are equipped with a Swiss manufactured 1951 model MG. and 8.05 cm 1951 smoke dischargers. This tank carries standard Swiss markings, a serial number consisting of a small white cross on red circle, [the Swiss national emblem] followed by the letter "M" and 5 digits, probably white on a black rectangle. Next to the serial is identification plate, with black stencilled lettering, 3 letters and a number arranged in a square. Non-standard is the three digit tac number in light color on hull rear and turret side.

[Above Right] Swiss Panzer 55/60 [Centurion Mk 5] with 105 mm gun. The spare road wheel is visible on left of driver. Note Swiss Model 1951 MG and 8.05 cm 1951 smoke dischargers. [K & W]

Swiss Panzer 57/60 [Centurion Mk 7 with Vickers L7A1 105 mm gun] equipped with Simfire on cross-country operation. [Solartron Schlumberger]

Swedish Stridsvagn 102. In 1962, Centurion Mk 3 and Mk 5 designated Stridsvagn 81 were modified to incorporate a 105 mm gun with the designation Stridsvagn 102, so as to match up with the Stridsvagn 101 [Centurion Mk 10], the primary distinguishing feature being the difference in mantlet. [Swedish Army]

[Above Right] Three Stridsvagn 102 moving across a Swedish field. [Author's collection]

Rear View of Bargninsbanvagn 81 [Centurion ARV Mk 2] of the Swedish Army. The lifting jib dismantled in two sections is fixed to the rear of hull. These two sections are joined together and erected on front of hull thus providing a 10-ton lift. Note also the dismantled 'A' frame. [Author's collection]

[Below & Below Right] Swedish Centurions usually tow a monowheel 910 litre armoured fuel trailer which can be discarded by firing two small explosive charges. [Author's collection]

A Danish Centurion Mk 5/2 equipped with Vickers L7A1 105 mm gun. Infantry tank battalions unlike tank battalions are equipped with Centurion Mk 5 retaining the original 20 pdr. gun. Both these versions have been modified to suit local requirements. These include German Mauser M-62 MG mounted coaxially with the main gun and an American 12.7 mm MG on commander's cupola. Markings are mainly British style, note convoy sign on deflector and triangular tactical sign. Serial number is in yellow repeated on each side of turret and on lower bow and rear plates. At these latter locations the number is usually preceeded by the Danish Hussar's crest, also in yellow. 120 Leopard 1A3's will soon be replacing the Centurion. [Author's collection]

Detail views of modifications to a Centurion Mk 5 of the Netherlands Army before these modifications were incorporated in the majority of Centurions still in service. Many of these Centurions incorporating some of the modifications such as Diesel engine, fuel tanks, cooling and exhaust systems have been exported to Israel. The new tracks, thermal sleeve on a Vickers 105 mm L7A1, increased volume of rear hull, IR searchlight are shown. Many Centurions of the Netherlands Army are fitted with new tracks equipped with rubber pads. These pads reduce noise when running on roads as well as increasing track life and giving a softer ride. [Herkenning Magazine]

[Above & Left] Netherlands Army Centurion with turret basket, RMG, 105 mm gun and thermal sleeve. 12.7 mm or 7.62 mm MGs are usually carried on the commander's cupola. A further modification which can be seen in both illustrations is the IR searchlight similar to the one fitted on the Leopard tank also in service with the Netherlands Army. Bridge classification disc and serial number are both black on yellow. [Herkenning Magazine]

Another Netherlands Army Centurion with rubber pads fitted to the tracks. [RAC Tank Museum]

Centurions advancing under cover of an Australian Bell helicopter. Close contact between helicopters and ground forces was maintained in Vietnam both for reconnaissance and supply of fuel invariably delivered by medium-lift helicopters in "seal-drums". One of the Centurions is fitted with an IR search-light. Once contact was made with the enemy the white light was often preferred to enable infantry to employ aimed fire. Following is a Centurion ARV Mk 2. On the anti-bazooka plate of the right-hand tank are the remnants of a tac sign that appears to have been applied with tape. [Bell Helicopter via Plaistow Pictorial]

Australian Centurion ARV Mk 2 shortly after arrival in South Vietnam. [Australian Army]

[Above Right] Centurion ARV Mk 2 pushes unserviceable Centurion aboard an LSM. [Australian Army]

U.S. built M579 Fitters Vehicle of the Royal Australian Electrical and Mechanical Engineers Light Aid Detachment delivering a vital piece of machinery to a Centurion Mk 5, which carries a British style Arm of Service mark, red over yellow with white numbering on the left hull front. [Australian Army]

Two recently-arrived Centurions in Vietnam fitted with dozer blades to clear suspected mined and booby-trapped areas. The side-plates had not yet been removed from Centurions. Close contact between armoured elements and helicopters was essential, especially at night when the Centurion's RMG proved invaluable as a target indicator to helicopter gun-ship pilots. [Australian Army]

A Centurion Mk 5 advances through thick jungle with commander ready to fire. The left track-guard has been torn by thick vegetation. Since gun ranges were too short, the special cap remained in position on the 20 pdr to protect gun. [Australian Army]

Amoured forces equipped with Centurion Mk 5s meet infantry equipped with M113s during operation 'Blue Mountains' in July 1968 to check enemy infiltrations. Note the 100-gallon tank bolted to the rear of hull and jerry cans stowed in turret basket. These fittings doubled the operating range of Centurion. Sideplates were soon discarded in South Vietnam due to build up of mud and vegetation causing distortion and damage to track-guards and track-guard bins. Furthermore, road wheels and suspension stations proved adequate in absorbing RPG2 attacks without hull penetration. [Australian Army]

Cambrai, famous World War I battle, celebrated as the birthday of the Armoured Corps, was commemorated in South Vietnam. A 100-gallon tank is bolted to the rear of this Mk 5, the turret basket rack has been adapted to mount the IR searchlight. The .30 in. ammunition is stowed on turret roof to give the commander a ready store. The tactical sign "1B" is repeated twice on the rear of the right hand tank, on the gas tank and on a loosely attached metal flap on the turret basket. "1A" is visible on the bore evacuator on the left hand tank. The chess knight, also in white, and a barely visible six digit serial number above it complete the markings. [Australian Army]

Infantry elements meet elements of the 1st Royal Australian Armoured Regiment. Cupola MG is US .30 cal. A name, probably a personal marking, has been painted on the 20 pdr barrel. [Australian Army]

Heat was the toughest enemy in the tropics. The crew of "Lolita" hit upon an ingenious way of protecting themselves from the sun by rigging up a hasty shelter over their turret. Jerry cans are stored at the rear of hull. The name of this tank has been delicately painted or chalked on the bore evacuator. [Australian Army]

An Australian Mk 5 of C Squadron. 1st Armoured Regiment, battering its way through the jungle on the border of Phuoc Tuy and Long Khanh Provinces, South Vietnam. [Australian Army]

Though in the process of being replaced by the Chieftan, the Centurion is still used by elements of the British Army. Seen here are soldiers of the 2nd Battalion Scots Guards dismounting from tanks of the Hong Kong Garrison.